Paulo Marcos Ferreira Andrade
Edinei F. S. Andrade
Iolanda S. Oliveira

Reviewing Knowledge: Building Perceptions!

Paulo Marcos Ferreira Andrade
Edinei F. S. Andrade
Iolanda S. Oliveira

Reviewing Knowledge: Building Perceptions!

On the road to overcoming the challenges of contemporary education

ScienciaScripts

Imprint

Any brand names and product names mentioned in this book are subject to trademark, brand or patent protection and are trademarks or registered trademarks of their respective holders. The use of brand names, product names, common names, trade names, product descriptions etc. even without a particular marking in this work is in no way to be construed to mean that such names may be regarded as unrestricted in respect of trademark and brand protection legislation and could thus be used by anyone.

Cover image: www.ingimage.com

This book is a translation from the original published under ISBN 978-3-330-19722-0.

Publisher:
Sciencia Scripts
is a trademark of
Dodo Books Indian Ocean Ltd. and OmniScriptum S.R.L publishing group

120 High Road, East Finchley, London, N2 9ED, United Kingdom
Str. Armeneasca 28/1, office 1, Chisinau MD-2012, Republic of Moldova, Europe
Printed at: see last page
ISBN: 978-620-8-23623-6

REVIEWED KNOWLEDGE: BUILDING PERCEPTIONS!

PAULO MARCOS FERREIRA ANDRADE[1]

EDIENI FERREIRA DA SILVA ANDRADE[2]

IOLANDA SILVA OLIVEIRA[3]

1 **Paulo Marcos Ferreira Andrade** Graduating in Portuguese/Spanish Literature, UFMT/UAB, Barra do Bugres Pole. Degree in Pedagogy - UNEMAT-Caceres. Specialist in "Pedagogical Coordination" and "School Management" at UFMT-Cuiaba. Specialist in rural education at Faculdade AFIRMATIVO (2016) and currently a study advisor for PNSIC in the municipality of Barra do Bugres - MT - e-mail: prof.paulomarcos@hotmail.com.
2 **Edinei Ferreira da Silva Andrade** has a degree in Pedagogy from FAEL (2012), a specialist in Early Childhood Education with an emphasis on school inclusion from AFIRMATIVO (2014), and has been a teacher in the municipal public network of Bara do Bugres - Mt- for over ten years, with experience as a literacy teacher in the proposals of the national pact for literacy. e-mail: rofessoraedinei@hotmail.com.
3 **Iolanda Silva Oliveira** - Graduated in Pedagogy from UNIP (2017), specialist in Early Childhood Education and rural education from Faculdade Afirmativo de Cuiaba.

SUMMARY

INTRODUCTION

"Resenhado Saberes Construindo Perceppoes" is a collection of critical reviews that are essential for teacher training. The texts in this collection deal with important themes in academia. They are dialogues with theories and knowledge constructed by different authors with a view to contributing to the training of teachers in the present century and to the innovation of their pedagogical praxis.

In this work, we bring to the focus of the dialogue texts such as: "Beyond Abyssal Thinking: From Global Lines to an Ecology of Knowledges" by Professor Boaventura de Souza; notes on "Conscientization: Theory and Practice of Liberation: An introduction to the thought of Paulo Freire; Edgar Morin's "Seven knowledges necessary for the education of the future" and finally a contextualization of teacher training: "Teacher Training in the Context of the Pedagogical Proposals of Rudolf Steiner (Waldorf Pedagogy), Maria Montessori and the Bridge School Experience" by Evelaine C. dos Santos. dos Santos.

Innovation is possible through the reprocessing of knowledge and at many times this also requires the deconstruction of ideas. Deconstructing ideas is a difficult process because they are culturally and socially ingrained in people's lives. However, this is a necessary process, but it is impossible without revisiting knowledge built on serious research in education.

In this way, what we have in this work is the revisiting of a serious and coherent idea in the field of academia and mathematics teacher training, with a view to promoting the innovation of knowledge necessary for education for change. The dialogues presented here are the fruit of prolonged study and reflection in order to understand the real needs of mathematics teaching in contemporary schools.

CHAPTER I

A Look Beyond Abyssal Thinking: From Global Lines to Bonaventure's Ecology of Knowledges

> The era in which we live must be considered a time of transition between the paradigm of modern science and a new paradigm, the signs of whose emergence are accumulating, and which, for lack of a better name, I call postmodern science. (BOAVENTURA S. S. 2002, p.11)

According to Santo (2007) "Modern Western thinking is abyssal thinking" (p. 1). Since this is an invisible line, it is often imperceptible in the process of social relations. This modern thinking then has visible and invisible lines of distinction. These radical lines are based on the maxim of division, separating modern society into two worlds, two distinct universes which, in the author's opinion, are characterized as "on this side of the line" and "on the other side of the line" (p.01).

In this thought, one can see that there is a side of the line determined by "this side", where there is recognition and a strong spirit of subsistence dominates. "The other side" of the line is therefore the place of the incomprehensible, irrelevant, invisible and insistent. There is then an imaginary but very real barrier that sets cultural and ideological limits to the maintenance of the unbroken line, in the dominant view. "The fundamental characteristic of abyssal thinking is the impossibility of co-presence on both sides of the line" (Santo, 2007, p.01).

As Boaventura S. S. (2007) points out, modern society is expressively based on the tension between
"regulation and social emancipation". This thought is confirmed by Gomes' arguments (2012, p. 43) when he postulates that:

> the dominant western epistemology was built on the basis of the needs of

4

colonial domination and is based on the idea of abyssal thinking. This thinking operates by unilaterally defining lines that divide experiences, knowledge and social actors between those that are useful, intelligible and visible (those on the other side of the line) and those that are useless or dangerous, unintelligible, objects of suppression or oblivion (those on the other side of the line).

There is therefore an existential conflict in the perspective of maintaining the territory of coloniality, maintaining the abyssal feeling as the fruit of normality and not as the result of the actions of a society whose pillars ignore ethics, obscure diversity as a form of equity, and benefit from it. In this context, there is the dichotomy of regulation/emancipation and the dichotomy of appropriation/. "We live [in] a time of transition between the paradigm of modern science and a new paradigm, the signs of whose emergence are accumulating, and which, for lack of a better name, I call postmodern science". (BOAVENTURA S. S. 2002, p.11)

"Modern abyssal thinking stands out for its ability to produce and radicalize distinctions" (BOAVENTURA S. S. 2007, p.02), the combined result of which is an extra-sensory feeling of permanence of whoever is on this or that side of the line, no matter how radical the distinctions are. An almost irrefutable feeling of belonging is created, causing regulation, control and social order according to the dominant interests.

In order to maintain the line and make it inviolable, it is necessary for "this side" to dominate the invisibility of "the other side", or to be effectively regulated and distinguished by means of policies. For example, programs such as "education for all" enable access through equality rather than equity. It guarantees entry to educational institutions but not permanence. "The intensely visible distinctions that structure social reality on this side of the line are based on the invisibility of the distinctions between this and the other side of the line" (BOAVENTURA S. S. 2007, p.2).

What Boaventura S. S. (2007) postulates here is that this makes them realize the existence of two lines with distinct subsystems that operate differently and interdependently, but with "visible and invisible distinctions in such a way that the invisible ones become the foundation of the visible ones" (p.03). It is precisely here that modern science has achieved a monopoly on distinction, at the heart of which are epistemological disputes. The hallmark of this context is in fact the epistemological dispute which, although invisible and concrete, is an attenuated conflict between the scientific and the empirical.

So there is a regulation of the relationship of knowledge, albeit relative truths, but which at any moment can claim the status of superior, and a relationship through tensions. "These tensions between science, philosophy and theology have always been highly visible, but as I argue, they all take place on this side of the line." (BOAVENTURA S. S. 2007, p.2). What is evident is that the visibility of recognized knowledge sustains the invisibility of unrecognized knowledge.

> On the other side of the line, there is no real knowledge; there are beliefs, opinions, magic, idolatry, intuitive or subjective understandings, which, at best, can become objects or raw material for scientific inquiry. Thus, the visible line that separates science from its modern "others" is based on the invisible abyssal line that separates, on the one hand, science, philosophy and theology and, on the other, knowledge that is rendered incommensurable and incomprehensible because it does not obey either the scientific criteria of truth or that of the knowledge, recognized as alternative, of philosophy and theology. (BOAVENTURA S. S. 2007, p.4).

There is also the question of law in modern society, which on this side of the line is established by the legal and the illegal as "the only two relevant forms of existence before the law" (p. 04). What Bonaventure makes a point of emphasizing in his discourse is that there is a radical negation of everything that exists or could exist on the other side of the line, separating "the true from the false, the legal from the illegal" (ibid.), thus creating the stereotype of "without a fixed territorial location", i.e. outside the colonial zone.

Being outside the colonial zone therefore means being part of the illegal, illegitimate, refutable, invisible. What is found on the other side has no legal merit, since "everything that cannot be thought of in terms of true or false, legal or illegal" (p.04).

The larger understanding is that the abyssal lines have very concrete colonial origins, yet they dominate contemporary thinking.

The denial created by colonialism creates a radical, emasculating absence, because the thought that there is no jurisprudence on the other side allows domination to perpetuate and control modern sub-humanity. The human being in these terms has no chance of inclusion. "The denial of one part of humanity is sacrificial in that it constitutes the condition for the other part of humanity to assert itself as universal" (Bonaventure S. S. 2007, p.5).

According to Boaventura S. S.'s thesis (2007), this process of colonial exclusion is still very present in contemporary society and "modern Western thought continues to operate across abyssal lines that divide the human world from the sub-human" (05). Modern society categorically reproduces the models of exclusion established during colonization. It's as if the anti-values they represent "on this side of the line" were, or are, impregnated in the way we do science, organize society and provide services to the classes. In this way, as Boaventura S. S. (2007, p. 05) postulates, "the creation and at the same time the negation of the other side of the line are an integral part of hegemonic principles and practices".

Whether in the field of science or law, without distinction, the divisions are postponed by the abyssal lines in such a way as to effectively eliminate any possibilities found on the other side of the line. "In these abyssal conceptions of epistemology and legality, the universality of the tension between regulation and emancipation, applied on this side of the line, does not contradict the tension between appropriation and violence applied on the other side of the line" (BOAVENTURA S. S. 2007, p.7). It is therefore possible to state that the consolidation of violence in the legal and epistemological aspects takes different forms.

In short, according to Boaventura S. S. (2000), the relationship between appropriation and violence is very profound indeed. However, appropriation involves aspects such as: incorporation, co-optation and assimilation, while violence implies physical, material, cultural and human destruction.

> "In the field of knowledge, appropriation ranges from the use of local inhabitants as guides and local myths and ceremonies as conversation tools, to the plundering of indigenous knowledge about biodiversity, while violence is exercised through the prohibition of the use of their own languages in public spaces, the forced adoption of Christian names, the conversion and destruction of symbols and places of worship, and all forms of cultural and racial discrimination." (BOAVENTURA S. S. 2007, p.6).

The relationship of power, imposition and domination at the extremes of the abyssal lines is evident, as it deals directly with the extraction of values and the manipulation of law, whether in the juridical, cultural or civil fields, which manifests itself in reality through different forms of *apartheid* and forced assimilation. In this context, according to the author, efforts must be made to understand the matrix that differentiates human law from the law of things (objects), since the logic of violence only recognizes the rights of things. Without this understanding and recognition, it is not possible to think about emancipation and regulation, since this requires an understanding of the human on both sides of the line. This means refuting the process of objectification of human nature, as a rule.

> Why, in the last two centuries, has an epistemology dominated which has eliminated the cultural and political context of the production and reproduction of knowledge from epistemological reflection? What have been the consequences of this decontextualization? Are other epistemologies possible today? Answering these questions means rescuing epistemological models that were once disregarded by the epistemic sovereignty of science. This can lead to the revaluation of identities and cultures that were intentionally ignored by colonialism for centuries. This was responsible for imprinting a historical tradition of political and cultural domination, which subjected knowledge of the world, the meaning of life and social practices to its ethnocentric vision. (GOMES, 2012, p. 40)

In this context, the constant tension between emancipation/regulation in terms of coexistence with appropriation/violence is made explicit, and these truths remain unquestionable. In a very precise way, the abyssal lines continue to structure modern thought. The author's thesis is that "the metaphorical cartography of global lines survived the literal cartography of the amity lines that separated the Old World from the New World" (BoaventurA S. S. 2007, p.09). In this way, there is a strong interconnection between social and cognitive injustice. "To be successful, this struggle requires new thinking, post-abyssal thinking" (09).

BOAVENTURA S. S. (2000) states that the abyssal lines "historically, the global lines dividing the two sides have been shifting" (09) and in the last sixty years have suffered two shocks. "The first took place with the anti-colonial struggles and the independence processes of the former colonies [...] The second shake-up of the abyssal lines has been taking place since the 1970s and goes in the opposite direction". It is therefore necessary to understand that in the relationship of conflict and persistence between emancipation/regulation, while the other side reacts against exclusion, on this side the efforts are opposed to it. This often ends up increasing the tension at the extremes of the lines, because on the other side an untrue feeling of coexistence is created.

It develops on the basis of uprising efforts against subaltern cosmopolitanism". According to Boaventura S. S. (2000), the movement consists of the return of a new abyssal colonial, with three main paths: "that of the terrorist, that of the undocumented immigrant and that of the refugee", although this does not mean a consolidated presence in metropolitan spaces, but only a relationship with the powers of this space. So the most obvious thing is that the classic colonial is gradually giving way to the new colonial with "a clear division between the Old World and the New World, between the metropolitan and the colonial. The line has to be drawn short enough to guarantee security" (12). In fact, this does not mean breaking the lines, but shortening them, without discarding the idea of segregation. An example of this is the relationship between governments and members of the commandos who settle in the favelas. It consists of a relationship of interest that makes each person's place on the lines very clear, albeit at a smaller distance. "It implies the resurrection of forms of colonial government, both in metropolitan societies, now affecting the lives of ordinary citizens, and in societies previously subject to European colonialism" (SANTO, 2000, p.13), which represents the return of the colonizer, but in a less frightening way.

The lines now in this new colony are drawn in both a real and metaphorical sense, but it continues with the marking of borders, dividing and segregating people and thoughts, what Bonaventure S. S. (2000) calls social fascism. "Social fascism is the new form of the state of nature, and it proliferates in the shadow of the social contract in two forms: post-contractualism and pre-contractualism" (15).

Although there are five forms of fascism, he discusses three:

a. **The fascism of social apartheid:** which highlights the segregation of urban, metropolitan cartography, making a categorical division between the savage and the civilized zones;
b. **Contractual fascism:** which deals with the relationship of power exercised by the parties in civil society. Here, the acceptance of the contracting party's conditions, by the party that is inferior due to lack of options or even opportunities, is the fascist maxim;
c. **c) Territorial fascism**. This would be the occupation of the machinery (state) by the bourgeois elite that dominates capital, the main characteristics of which are the co-optation and violation of state institutions;

From the perspective of social fascism, Boaventura S. S. (2000) states that "it is therefore a pluralist fascism and, therefore, a form of fascism that has never existed" (16). This leads us to understand that society is entering "a period in which societies are politically democratic and socially fascist" (ibid.).

In short, modern abyssal thinking regulates the relationship between the citizen and the state, what Boaventura S. S. (2000) calls "modern domination". This makes it very clear that abyssal thinking has not been exhausted, nor has it given way to other thinking; on the contrary, it continues to be reproduced.

> This means that the critical task ahead cannot be limited to generating alternatives. In fact, it requires alternative thinking of alternatives. We need a new way of thinking, a post-abyssal way of thinking. Is this

This reflection brings us back to the establishment of post-abyssal thinking. According to Boaventura S. S., post-abyssal thinking is based on the idea that the diversity of the world is inexhaustible, continuous, and lacks an adequate epistemology. Considering four prominent aspects, namely:

1. post-abyssal thinking arises from the concept that the diversity of the world is inexhaustible, and that diversity needs an appropriate epistemology.
2. thought understands that social exclusion takes different forms, and as long as exclusion continues abysmally, no progressive post-capitalist alternative is allowed.
3. post-abyssal thought is a thought that is not derivative, because it consists of a break with western models and the condition for its existence is the presence of radical co-presence identifying the practices and agents on both sides of the line are contemporary in egalitarian issues.
4. Radical co-presentation abandons the linear conception of history, the denial of war and intolerance, leading us to a new way of understanding history, emphasizing that contemporaneity is simultaneity, understanding that agents and supporters are contemporary and equal (PAIVA, 2015, p.06).

Social exclusion must now be recognized in a broader sense, according to Boaventura S. S. (2000). There is therefore a categorical differentiation in the forms of exclusion along the lines established.

In this way, there is thinking about the exclusion that comes from a non-abyssal line and thinking about the abyssal persistence of exclusion. This really is very complex. Boaventura S. S. (2000) insists on the argument that "recognizing the persistence of abyssal thinking is a *sine qua non* condition for starting to think and act beyond it" (21). In other words, post-abyssal thinking can only be established through radical coexistence.

Thus, what marks the presence of post-abyssal thinking is the ecology of knowledges through the maximum inexhaustible epistemological diversity of the world. "The ecology of knowledges does not conceive of knowledge in the abstract, but rather as knowledge practices that enable or prevent certain interventions in the real world" (BOAVENTURA S. S.; MENESES, 2009, p 49).

Paiva (2015, p. 07), when analyzing Boaventura's propositions on the ecology of knowledges, postulates that "it is present as a means of intervening in reality and a dialogue in society, consolidating itself in a pragmatic and epistemological aspect". "The main challenge of the ecology of knowledges is the modern belief in science (whether as a belief or an idea) as one of the forms of valid knowledge, one of the main characteristics of abyssal thinking." (BOAVENTURA S., p. 1) (BOAVENTURA S. S. 2009 p 46)

In this way, it can be said that the ecology of knowledges is basically a counter epistemology. Thus, the political emergencies of peoples and resistance to global capitalism are the mainstays of the ecology of knowledges. What can be seen is that the ecology of knowledges is based on a framework that tends to value both cognition and ignorance, and does not establish boundaries of exclusion. Until now, the result of the abyssal lines has been the failure to recognize the scientific in an equitable way, which would only happen with the overcoming of global capitalism. The understanding we can have is of the destabilization caused by the ecology of knowledges in a radical critical movement, "of the politics of the possible, without giving in to an impossible politics" (BOAVENTURA S. S. 2000, p.30).

This leaves us with the understanding that this epistemological formation is something extremely difficult, yet possible. In conclusion, Boaventura identifies three main sets of issues surrounding this subject: "The first set refers to the identification of knowledges and raises a series of questions that have been ignored by the epistemologies of the global North." In this way, the idea of distinguishing between different types of knowledge and recognizing plurality is put forward. The second point addresses aspects of the procedures that interfere with this recognition and identification and finally questions the evaluation procedures of these interventions.

CONCLUSION

The discussion made possible by Boaventura S.S. (2000) in "Beyond Abyssal Thinking: From Global Lines to an Ecology of Knowledges" in fact provokes a rupture in modern thinking.

The maxim with which the author dialogues in his propositions makes us think about the paths taken by modern society, which for the most part only reproduces colonialism in a more camouflaged form and continues to make class, gender and ethnic distinctions, effectively contributing to the reification of humanity.

We need to break down the abyssal lines, not in the sense of shortening them, but in the sense of enabling cognitive, epistemological and cultural formation in an equitable way. Man needs to be human and not a thing, and society should not be seen as a machine that serves as an instrument for perpetuating distinctions between this side and the other. In short, the author's proposal is to break down these distinctions through the ecology of knowledge and by breaking down global capitalism.

References

BOAVENTURA S. S. **Beyond Abyssal Thinking: From global lines to an ecology of knowledges**. Revista Critica de Ciencias Sociais, 78, October 2007: 3-46

BOAVENTURA, S. S Meneses, M.P Epistemologies of the South. Coimbra. Almeidina, 2009.

Introduction to a postmodern science. Rio de Janeiro: Graal, 2002

A discourse on science. 9ª ed. Porto: Afrontamento,1997

Epistemologies of the South. Coimbra. Almeidina, 2009.

GOMES, Fulvio de M. **As Epistemologias do Sul de Boaventura de Sousa BOAVENTURA S. S. : por um resgate do sul global**. Revista Paginas de Filosofia, v.4, p, 39-54, dec 2012.

PAIVA. Marilia Luana Pinheiro de. **A Look at Boaventura de Sousa's "Epistemologies of the South" BOAVENTURA S. S.** . Uniara Magazine. Volume 18, n° 1, July 2015

Chapter II

**Revisiting the Concept of Conscientization: Theory and Practice of
Liberation: An Introduction to the Thought of Paulo Freire**

The ideal I'm fighting for requires me to create in myself the will to fight
alongside the courage to love.

Paulo Freire

When we take Freirian thought into the dialogue on conscientization, we assume that this is existentialist pedagogy. In order to understand the concept of conscientization that he puts forward, it is necessary to understand that man is the subject of the relationships he establishes with the world. It is therefore understandable that the relationships that man establishes with the environment have significant dimensions in the process that distinguishes him as a rational being. In this way, it is from the concept and meanings of plurality, criticality, temporality and consequences that Freire establishes his thinking on conscientization.

In philosophical terms, Western thought has taken a somewhat clear stance on the relationships established by man. In these terms, we can see an idealist vision that has the essential self as its primacy. There is also objectivism, based on uncriticism and mechanism, which denies man as an active subject. However, Freirian thought does not conceive of consciousness in any of these extremes, but rather in the dialectical interplay of man-world relations.

So reality may not perceive man as a cognizant being in the first instance because of his critical awareness, but because of his naïve position.

Freire (1979, p.15) states that:

> In this sense, awareness is a reality check. The more we become aware, the more we "unveil" reality, the more we penetrate the phenomenal essence of the object we are facing in order to analyze it. For this very reason, awareness does not consist of "standing in front

Based on this fragment from Freire, man must be understood as a historical being capable of creating and recreating the world. In view of this, he argues that "conscientization is not based on consciousness on the one hand and the world on the other; on the other hand, it does not intend separation. On the contrary, it is based on the relationship between consciousness and the world" (FREIRE1979, p.15).

In this context, we have the idea that awareness is based primarily on the critical development of a state of spontaneous consciousness. At the same time, it permeates reality itself, with cognizant action as its object, in which the subject takes on an epistemological role. This is not merely apparent, but a process that generates the refutation of false consciousness and naïve consciousness. Frei then proposes that from this demystified awareness, man can achieve a more effective insertion into the environment in which he lives.

In this way, the process of political literacy - like the linguistic process - can be a practice for the "domestication of men", or a practice for their liberation. In the first case, the practice of conscientization is not possible at all, while in the second case the process is itself conscientization. Hence a dehumanizing action, on the one hand, and an effort towards humanization, on the other. (FREIRE1979, p.16).

The link between utopia and conscientization is evident, in a process in which the conscientized man becomes a historical denouncer, while taking on personal and collective transformation.

This utopia mentioned by Freire does not refer to something unrealizable; on the contrary, it is what he calls the dialectic of the acts of proclamation and denunciation. It is essential that the production of knowledge is in this way intimately consubstantiated between consciousness and reality. "The dialectical vision rejects the understanding of consciousness as a pure reflection of material objectivity, and also demonstrates the incompatibility between it and the idea of an inexorable tomorrow" (FREIRE, 2000). This coming to consciousness by the subject is in fact his own confrontation with reality.

The insertion of the subject into reality as a conscious subject causes the preponderant knowledge to be overcome, but it is necessary for the subject to perceive reality in its entirety. Conscientization is therefore the process that distances the oppressed from the oppressor and any and all situations of oppression. It has to be considered in this way that "conscientization produces demythologization. It is obvious and impressive, but the oppressors will never be able to provoke awareness for liberation: how can I demythologize if I oppress?" (FREIRE1979, p.16).

The work of humanization can only be the work of demythologization. This is why conscientization is the most critical look possible at reality, which "unveils" it in order to get to know it and to get to know the myths that deceive and help maintain the reality of the dominant structure. (FREIRE1979, p.16).

Thus, the total perception of reality is a crucial principle for the process of conscientization from Freire's perspective. In other words, the basic principle of education is awareness, which is an immediate operation of the active subjects. In this way, while acting, the subject regains awareness and transformation as a result of a profound process of change. Conscientization thus operates in the subject while they act and work. In this sense, Freire (2001, p. 36) postulates that "... knowledge does not extend from those who believe they know to those who believe they do not know;
knowledge is constituted in human-world relations, relations of transformation, and is perfected in the critical problematization of these relations".

This is undoubtedly awareness as a human endeavor carried out through human-nature mediation. In these terms, the subject critically takes possession of awareness, while also occupying the world through the revealing actions of consciousness. Thus, as subjects of transformation, they also humanize themselves, since this is not a dissociable process in practice (FREIRE 2001).

"This critical appropriation pushes them to assume their true role as men. That of being subjects of the transformation of the world, with which they humanize themselves" (p. 36).

This is not a process that Freire conceives in an imposed way, since this would not result in awareness. On the contrary, he posits that teacher and pupils are the subjects of a solidary and conceptualizing literacy that gives its subjects choices. Whereas "... in the educational practice of the progressive option, it will never cease to be an unveiling adventure, an experience of unveiling the truth" (FREIRE 2000, p. 9).

Freire's maxim is that conscientization is the fruit of the unveiling of truth, and that this happens through cognitive construction. "Conscientization invites us to assume a utopian position vis-à-vis the world, a position which converts the person made conscious into a 'utopian factor'" (FREIRE, 1979, p.16).

> There's no reason not to repeat that teaching isn't just a mechanical transfer of the content profile from the teacher to the student, who is passive and docile. Nor can we repeat that starting from the knowledge that the students have does not mean revolving around this knowledge. To start means to set out, to go, to move from one point to another and not to stay. I have never said, as people sometimes suggest or say I have, that we should spin around the students' knowledge like a moth around light. To start from the 'knowledge of experience' in order to overcome it is not to remain in it (FREIRE, 2000, p. 70 and 71).

Awareness is, in fact, this determined flight to overcome what has been done. To know is not to focus on the beginning, it is to go beyond, to set out; the opposite of this is to die in the light with a glimpse of darkness. According to Freire (2001, p.83) "it is the problematization of the world of work, of works, of products, of ideas, of convictions, of aspirations, of myths, of art, of science, in short, the world of culture and history, which, resulting from human-world relations, conditions men themselves, their creators" that generates awareness from a perspective of launching and not of stagnation. For Freire, the proposition of 'ad- mirem' is necessary, i.e. that man critically views the world in a totalized way, the result of which is action on the world itself, transforming the space but also man (FREIRE, 2001).

It's about students and teachers being able to take school learning into their world, the world of the community in which they live. And to use it as a way of communicating and broadening their reading of the world. Problematizing it and themselves in a continuous and dialectical process. In this way, "when men perceive reality as dense, impenetrable and enveloping, it is indispensable to proceed with this search by means of abstraction (...) if the two elements, as opposites, are maintained in dialectical interrelation in the act of reflection" (FREIRE, 1979, p.17).

In this context, the different investigative processes of conscientization education become evident, while a number of principles are built up:

1. To be valid, every education, every educational action must necessarily be preceded by a reflection on man and an analysis of the concrete living environment of the concrete man whom we want to educate (FREIRE, 1979, p.19).

This process of reflection on man is extremely important, despite the idolization of methods. In this way, it is man who thinks about the method and not the method that determines man. "Man's vocation is to be a subject and not an object. In the absence of an analysis of the cultural environment, one runs the risk of carrying out a prefabricated education, which is therefore ineffective and not adapted to the concrete man for whom it is intended" (FREIRE, 1979, p.19). This understanding is based on the fact that method is manipulable.

2. Man becomes a subject by reflecting on his situation, on his concrete environment, (FREIRE, 1979, p.19).

This is what Freire calls the process of humanization that provokes adherence, while the latter means perceiving and moving away from the situation of the object, the oppressed, and seeking liberation and humanization. The greater the reflection on reality, the more fully, concretely and committed the action will be. In Freirian thinking, "education that seeks to develop the awareness and a critical attitude, thanks to which man chooses and decides, frees it instead of subjugating it, taming it, adapting it" (p.19).

3.   To the extent that man is integrated into his context,
reflects on this context and commits himself, constructs himself
himself and becomes a subject. This guiding idea can be separated into two
statements (FREIRE, 1979, p.20).

The truth is that man, as a thinking and critical being, is capable of recognizing and recognizing himself in realities." This ability to discern what is not proper to man also allows him to discover the existence of a God and to establish relationships with him" (FREIRE, 1979, p.20). In this way, he is referred to the flow of becoming, where he is trapped in an eternal present, on the contrary, in continuous progression. Man will become a subject through the relationship he makes with the objective and the subjective, with the reflexive relationship of the environment in which he lives. He becomes a subject "in the very act of responding to the challenges presented to him by his context of life, man creates himself, realizes himself as a subject, because this response demands of him reflection, criticism, invention, election, decision, organization, action" (FREIRE, 1979, p.21).

4.   To the extent that man, integrating himself into the conditions of his life context, reflects on them and comes up with answers to the challenges presented to him, he creates culture (FREIRE, 1979, p.21).

"From the relationships he establishes with his world, man, creating, recreating, deciding, dynamizes this world. He contributes something of which he is the author... By this fact he creates culture" (p.21). This establishes the appropriation of culture as a significant and systematic element, which, unlike nature, is the result of man's own actions. This is configured as a critical and systematic acquisition, the more culture is produced the more criticality will be possible, "to say that man cultivates and creates culture in the act of establishing relationships, in the act of responding to the challenges presented to him by nature..." (ibid.)

5. Not only through his relationships and his responses is man the creator of culture, he is also the "maker" of history. To the extent that human beings create and decide, epochs are formed and reformed (FREIRE, 1979, p.21).

It is clear that man produces history and that the construction of individual and collective memories is also necessary, but we need to look at the history of the people and not just that of the rulers. It is therefore necessary to allow ourselves other angles without which awareness will not be raised. For according to Freire (1979, p. 22) "man cannot participate actively in history, in society, in the transformation of reality, if he is not helped to become aware of reality and of his own capacity to transform it".

One of the premises of Freire's reflection is the universe in which the method is applied. Since the method must be manipulated by man, who, as already mentioned, is not a passive object, its application depends on knowledge of the immediate reality. This implies both the socio-affective-economic context and the cultural and vocabulary context. Freire's inference is of a method that gives wings and not cages, "an instrument of the learner, and not only of the educator, and which identifies - as a Brazilian sociologist rightly pointed out - the content of learning with the very process of learning" (FREIRE, 1979, p.21).

In this context, Freire ends up breaking down the application of the method into five distinct phases:

1. **First phase:** the "discovery of the universe of vocabulary", which directly implies the start of the research and delving into the reality we are trying to reach;
2. **Second phase:** Selection of words, within the vocabulary universe, which he attributes to three indispensable elements: "a) Syllabic richness; b) Phonetic difficulties. b) The practical content of the word, which implies seeking the greatest possible commitment of the word in a de facto, social, cultural, political reality" (FREIRE, 1979, p.21);
3. **Third phase:** The third phase is the creation of existential situations typical of the group you are working with;
4. **Fourth phase:** The fourth phase is the preparation of index cards to

help the debate coordinators in their work.

5. **Fifth phase:** This consists of drawing up cards on which the phonetic families corresponding to the generator words appear.

This literacy process inferred by Freire seeks awareness and cognitive construction in a dynamic and inseparable process. It consists of a dialogic process where educator and student build together and learn together.

Freire (1979, p. 72) comments that:

> Literacy cannot be done from the top down, nor from the outside in, like a donation or an exhibition, but from the inside out by the illiterate themselves, only adjusted by the educator. That's why we were looking for a method that could also make an instrument of the learner and not only of the educator.

The method developed by Freire presents the educated with a great challenge, namely to teach literacy beyond the primer, i.e. to manipulate knowledge so that it is not a mechanical act that does not result in awareness but only in memorization.

He points out that:

> [...] it would be impossible to engage in mechanical memorization of ba-bebi-bo-bu, la-le-li-lo-lu. That's why I couldn't reduce literacy to the pure teaching of words, syllables or letters. Teaching in which the literacy teacher "fills" the supposedly "empty" heads of the literate with their words. On the contrary, as an act of knowledge and a creative act, the literacy process has the literate as its subject. The fact that they need the educator's help, as happens in any pedagogical relationship, does not mean that the educator's help should cancel out their creativity and their responsibility in constructing their written language and reading this language. In fact, both the literacy teacher and the child, when they pick up, for example, an object, such as the one I have between my fingers, feel the object, perceive the felt object and are able to verbally express the felt and perceived object. [...] Literacy is the creation or assembly of written expression from oral expression. This assembly cannot be done by the teacher for or on the literacy student. There he has a moment of his creative task. (FREIRE 1989, p.13)

Taking up Marxist thought, Freire gives man himself the task of transforming reality, which is possible through a pedagogical praxis for liberation. For man only transforms if he is liberated, and for this he needs to perceive oppression.

The pedagogy of the oppressed, as a humanist and liberating pedagogy, will have two distinct moments. The first, in which the oppressed unveil the world of oppression and commit themselves in praxis to its transformation; the second, in which, once the oppressive reality has been transformed, this pedagogy ceases to be the pedagogy of the oppressed and becomes the pedagogy of men in a process of permanent liberation. (FREIRE, 2005, p. 46)

What is actually proposed is an understanding that the praxis of liberation must promote the dialogicity inherent in educational practice. Thus, the praxis of liberation must return to the dialogic matrix to the point of allowing cultural and human reflection on itself and on man. "The revolution that transforms a concrete situation of oppression, by linking the process of liberation, must still face this phenomenon" of dialogicity, (FREIRE, 1979, p.31). Despite the lack of dialogicity, "many of the oppressed who participate directly or indirectly in the revolution, conditioned by the myth of the old order, seek to make it their own. The shadow of their former oppressor is continually cast over them (ibid).

The praxis of freedom does not seek the repetition of the action of oppression by the oppressed. This would not be liberation, but a reversal of position, although this is the principle of education today. Transformation towards freedom must rely on man as an agent for transforming reality. "The true humanist recognizes himself more through trust in the men who lead him to commit himself to a struggle than in the thousands of supports he can undertake for them, without this trust" (FREIRE, 1979, p.32).

In short, the process of awareness requires a change of concept and a more affectionate belief in the human person and in the process of humanization. Thus, having the critical awareness that one must be the owner of one's work and that "this constitutes a part of the human person" and that the "human person cannot be sold or sell itself" is a step beyond palliative and misleading solutions. It is to be part of a real transformation of reality so that, by humanizing it, men can be humanized. (FREIRE, 2005, p. 212)

CONCLUSION

Awareness, according to Freirean thinking, goes beyond the act of constructing knowledge. It is the implication of this knowledge in a process of critical and continuous reflection on the relationship between man and nature. There can be no conscientization without reflection and without recognizing that man can transform the environment in which he lives as he produces knowledge.

> To the extent, however, that men's consciousness is conditioned by reality, conscientization is, first and foremost, an effort to rid men of the obstacles that prevent them from having a clear perception of reality. In this sense, conscientization produces the repulsion of cultural myths that alter men's consciousness and turn them into ambiguous beings. (FREIRE1979, p48).

According to Paulo Freire, liberatory praxis is the development of the conditions necessary for the continuous and gradual exercise of freedom. Not the reproductive freedom of oppression, but the freedom of socio-cultural-economic transformation. Educating for awareness is therefore from this perspective "undertaking revolutions to liberate men, precisely because men can know that they are oppressed and be aware of the oppressive reality in which they live" (FREIRE1979, p48).

REFERENCES

BRANDAO, C. R. **Historia do menino que lia o mundo**. 2.ed. Veranopolis, RS: ITERRA, 2001

FREIRE, Paulo. **Conscientización: teoria** e pratica **da libertação: uma introdução ao pensamento** de **Paulo Freire** [translated by Katia de Mello e silva; technical review by Benedito Eliseu Leite Cintra]. - Sao Paulo: Cortez & Moraes, 1979.

Pedagogy of Autonomy: knowledge necessary for educational practice. Sao Paulo: Paz e Terra, 1999a.

Pedagogy of Hope: a re-encounter with the Pedagogy of the Oppressed. Sao Paulo: Paz e Terra, 2000.

Pedagogy of the oppressed. 45. ed. Rio de Janeiro: Paz e Terra, 2005.

Extension or Communication? Rio de Janeiro: Paz e Terra, 2001.

The importance of reading: in three complementary articles.
23.ed. Sao Paulo. Associated authors: Cortez, 1989

CHAPTER III

A Brief Reflection on Edgar Morin's Seven Knowledges Necessary for the Education of the Future

The Seven Knowledges Necessary for the Education of the Future is the result of the author's commitment and effort in response to a request from UNESCO in 1999. The book then systematizes a set of seven knowledges or ideas that serve as a springboard for thinking about education in the present century.

The book is structured in seven chapters chronologically entitled as follows: the blindness of knowledge: error and illusion; the principles of relevant knowledge; teaching the human condition; teaching earthly identity; facing uncertainty; teaching understanding; the ethics of the human race.

Each chapter contains Morin's main ideas for rethinking 21st century education. The author effectively addresses the reflection needed to rethink contemporary education, using guiding axes to outline the paths to be taken. In other words, Morin presents arrows on the road to the necessary rethinking. These arrows he calls "The Seven Knowledges Necessary for the Education of the Future". The book is written in accessible language and contains the elements needed to revise pedagogical praxis, pointing to a more effective inclusion of the human aspect in cognitive construction.

Each chapter offers a possibility for change and puts old traditions on the back burner, refocusing the theoretical view while questioning the foundations of educational practices. The arrows Morin points out in Seven Knowledges emphasize the solidary fraction of education in a particular way, while urging the educators of the future to redefine their educational precepts.

This reflection systematically follows the chronological order of the chapters presented by the author. It thus alludes to his proposals and makes critical reflections on each axis presented, consolidating them with theories of contemporary education.

III.   I. The blind spots of knowledge: error and illusion

> All knowledge carries the risk of error and illusion. [...] Education must
> show that there is no knowledge that is not, to some degree,
> threatened by error and illusion. [...] Knowledge is not a mirror of things
> or the external world. All perceptions are, at the same time, brain
> translations and reconstructions based on stimuli or signals captured
> and coded by the senses. (MORIN, 2000, p.19-20)

According to Morin's propositions, cognitive processes are permeated by a sudden blindness to what is human. In this way, it is evident in the discourse that the educational process is connected to the service of capitalism, where the most important thing is to do, not to make people know what it is to know. And so it ends up leading to the error of illusion.

In view of this, the author works with the construction of the skill of knowing knowledge as a necessity that precedes the reality of being "ready-made". On the contrary, it is necessary to know the paths, the methods and the final product, which in education will never be finished but will always be an arrow pointing to a new path to be followed.

Knowledge of knowledge, in other words, is preparation for facing the permanent risks of error and illusion that constantly parasitize the human mind. So what Morin proposes is that educational processes first arm the human mind with knowledge, making it known to the point where the human mind proceeds vitally in a state of lucidity. The education of the future must therefore be aligned with a process that is both psychic and cultural, which encompasses the dimensions of the human being and the production of knowledge.

According to the author's thinking, education is a process of becoming aware, where there is no knowledge that is not threatened by the error of illusion. In short, what needs to be understood is that as a primarily mental process, the production of knowledge is not a mirror of the outside world, and is therefore made up of perceptions, beliefs, values and mental reconstructions. Mental reconstructions are based on the signals captured by the senses and these are not always reliable.

Perhaps we should reflect on the weaknesses of knowledge production processes that are based on a socio-cultural reality, and abstract knowledge from the senses alone without confronting it with the immediate context.

According to Morin (2000, p. 20) "No cerebral device allows us to distinguish hallucination from perception, dreams from vigilance, the imaginary from the real, the subjective from the objective". However, the author stresses the extreme importance of these mental processes which connect the exterior, represented by a percentage of 2%, with the interior, 98%.

This context inferred by Morin allows us to understand that knowledge as an idea, or even materialized in words, will always be the result of a translation/construction relationship through the psychic processes of language and the formation of human thought. In these terms, it can be said that the author postpones and that this knowledge is undoubtedly subject to error. This is because knowledge allows for the possibility of interpretations. It is precisely the possibilities of interpretation to which knowledge is exposed that generates the risk of error as it works on subjectivity.

Reflection on the question of error is indeed necessary, since these are the result of man's own rational controls. Elements such as emotions, fear, beliefs and desires are disturbances that considerably increase the risk of error (MORIN, 2000). It is in this perspective that a maxim postulated by Morin is found, namely that there is no accuracy that frees knowledge from the risk of error, while thinking the opposite is the greatest error, that is, underestimating illusion is the greatest illusion.

This calls for a more acute perception of the cognitive process, since the act of communicating knowledge may not actually mean making it known to a knower. It is therefore important to understand that perception is subject to some noise or intervention from random external processes.

In this context, Morin's (2000) proposition about the effectiveness of scientific knowledge as a weapon in the perception of errors and illusions becomes clear. However, there are controlling paradigms of science that can produce illusions because no scientific method or theory is free or immune from error. The fact is that scientific knowledge has not yet succeeded, and perhaps never will, in encompassing epistemological, philosophical and ethical paradigms.

In this regard, the primary task of education would be to identify the origins of errors, illusions and blindness, with this as necessary knowledge and as an arrow in the road to overcoming them.

According to Morin (2000) it is possible to detect:

Mental errors - Memory itself is a source of countless errors.

Intellectual errors - theories, doctrines, ideologies - are not only prone to error, they also protect and perpetuate it.

The errors of reason: although rationality presents itself as the best and most effective form of protection against error, it can still bring with it the risk of illusion and error.

The education of the future, in these terms, will not be the result of science or human rationality; on the contrary, it will emerge from rational uncertainty, constant risks and extreme self-criticism. Morin (2000, p.24) states that: "The game of truth and error is not only played through empirical verification and the logical coherence of theories. It is also played, deeply, in the invisible zone of paradigms. Education must take this into account". In this direction, the author works on understanding the blindness of knowledge.

So he takes a systematic approach:

    a. Promotion/selection of the master concepts of intelligibility;
    b. Determination of the master logical operations. The paradigm is hidden under the logic and selects the logical operations that become at the same time preponderant, pertinent and evident under its dominion;

Paradigmatic errors - explanatory models - paradigms - are also subject to errors - of conception and interpretation of concepts. The Cartesian paradigm, for example - the mainspring of Western scientific and cultural development - is based on binary contrasts: subject/object, soul/body, spirit/matter, quality/quantity, feeling/reason, existence/essence, right/wrong, beautiful/ugly, etc. - do not find in today's world the foundation they seemed to have at the beginning of the 20th century. The paradigm - like the Cartesian paradigm - shows something and hides something else - and can therefore elucidate and blind, reveal and conceal. It is within it that the problem is hidden - the key to the game of truth and error.

What is proposed is the dissociation of the extremes: Subject/Object, Soul/Body, Spirit/Matter, Quality/Quantity, Purpose/Causality, Feeling/Reason, Freedom/Determinism, Existence/Essence.

Morin also works with the concepts of imprinting and normalization. Imprinting can be said to be the often imperceptible marks of the first human experiences as an animal, i.e. the link to the culture built by the family cycle. This cultural imprinting marks humans at every moment of their lives, including the process of education. Normalization is related to the standardization of consciences. Normalization is strongly marked by complacency and the elimination of the ability to create and recreate life, which in this case would be the cultural imprinting itself.

From this context, Morin uses the considerations of Marx, who said: "The products of the human brain have the appearance of independent beings, endowed with particular bodies in communication with humans and with each other". We can see that the author uses this quote to highlight the process of embodiment, where beliefs and ideas are merely products of the human mind.

Given this, it's possible to say that man is, in short, captive to his own idea or perception of the world. In other words, his beliefs create an intrinsic form of persuasion as pressure and control of ideas, what Morim calls noology. However, in the face of the security and conformism of our beliefs and ideas, the new can emerge and break paradigms, causing a process of deconstruction, often necessary, which the author calls unexpected.

In fact, the unexpected doesn't always materialize comfortably since it requires paradigm shifts, rethinking and reprocessing beliefs and ideas. The unexpected generates uncertainties in knowledge, while it leads to deconstruction and construction. Man does not challenge himself very often, or almost never, to the unknown as a possibility of knowing. Morin highlights the great questions contained in the process of knowing, since knowledge itself is driven by questions. Seeking to know is an indispensable process for which education must provide the necessary pillars.

Putting questions into practice is the oxygen of any knowledge proposal. And knowledge remains an adventure for which education must provide indispensable support.

III.II The principles of relevant knowledge

> Knowledge of key problems, of key information about the world, however random and difficult it may be, must be attempted under pain of cognitive imperfection, even more so when the current context of any political, economic, anthropological, ecological knowledge... is the world itself (MORIN, 2000, p.33).

According to Morin's propositions, there is always a key problem, which is categorically configured as the emerging need to produce knowledge capable of making it possible to understand global problems. We are still dealing here with the vicissitude of knowledge and the act of knowing, but now knowledge is characterized as a tool for demystifying the problem.

This, of course, will be possible when there are no more ruptures in knowledge, i.e. fragmentations that prevent knowledge of the whole, and emphasize apart as essential. In Morin's opinion, in order to understand the key or global problem, it is extremely necessary to definitively replace fragmentation with the whole.

In this way, the processes of building knowledge are based on complexity. However, this requires a break with the old system of fragmentation, situating the process not in the part but in the whole, in a complex context, which he assures us is a skill that must necessarily be built. In this way, we understand that "the relevance of the world as a world is both an intellectual and a vital necessity".

Morin also argues that knowledge in this context is recognizing and getting to know the world's problems. However, for this to become a reality, he returns to the question of the unexpected, that is, the need for change, for engagement with the new. This would result in the organization of knowledge. For the author, the education of the future must necessarily face up to its paradigm, which is to make the following elements visible in its knowledge-building process:

a. **context**: knowledge of information or data in isolation is insufficient; information and data must be placed in their context in order for them to acquire meaning. Bastien, quoted by Morin (2000, p. 34), emphasizes that "cognitive evolution is not moving towards the establishment of increasingly abstract knowledge, but, on the contrary, towards its contextualization";

b. **global:** as a process that deals with the whole, it is much more than one and more than the context, since it is the relationship of various parts that operate in the organizing whole of which we are a part;

c. **multidimensional:** the human being or society is multidimensional and at the same time biological, psychological, social, affective and rational;

d. The complex: relevant knowledge must deal with complexity as the union between unity and multiplicity.

The considerable advances of the era in which we find ourselves, which Morin calls planetary, are drastically pushing us towards the challenges of complexity. The momentum of this planetary era highlights the need to develop a general intelligence in order to improve specialized skills and competences.

What the author emphasizes is not only the need for a general intelligence, but that education should articulate the elements necessary to build this intelligence. Morin (2000) argues that it is not necessary to annihilate the idea of discipline, but to rearticulate the idea of discipline in other contexts. And this requires the organization and activation of a body of knowledge, which would be possible with the instigation of curiosity as a tool. According to Morin (20000), the education of the future is intimately challenged to establish and promote general intelligence in subjects. And this implies identifying the false rationality that prevents the subject's progression towards general intelligence.

The author goes on to say that it is necessary to put an end to the antinomy that causes the dichotomy between the sciences and the humanities. Thus, the education of the future must stop being hyper-specialized, where each person is responsible only for their own specialized task, weakening global perception.

Morin identifies disjunction and closed specialization - hyper-specialization - as essential problems, since both are obstacles to global perception. There is what he calls the principle of reduction, which consists of a strong tendency to focus on the complex to the detriment of the simple. The fact is that education has made exactly this restriction because it has always strived to compartmentalize, isolate and not unite knowledge, all of which constitutes an unintelligible jigsaw puzzle.

This type of education breaks with the complexity of the world, combining it with the disjointedness that compartmentalizes knowledge, resulting in a myopic intelligence. The 20th century is marked by two characteristics: the first is its technological and scientific advances and the second is its blindness to global problems. This is the false rationality that formulates technocratic thinking.

For knowledge to be relevant, education must make the context, the global, the multidimensional and the complex evident. There are sciences that already practice relevant knowledge, such as ecology, which brings together different areas of knowledge. Therefore, relevant knowledge is an idea against fragmentation

III.III. Teaching the human condition

> The education of the future must be first and universal, centered on the human condition. We are in a planetary age; a common adventure leads human beings wherever they are. They must recognize themselves in their common humanity and at the same time acknowledge the cultural diversity inherent in all that is human. (MORIN 2000, p. 45)

Morin's understanding of man is in fact that of the human. This perception frees itself from myopia. In this way, the human being is at once physical, biological, psychological, cultural, social and historical. His complex nature manifests itself while education with disciplinary teaching struggles to disintegrate this nature.

In Morin's thinking, the ideal is for the human condition to become the object of each and every human process. In this way, education would indeed become complex, as humanity is, bringing together and organizing knowledge scattered across the natural sciences, the human sciences, literature and philosophy, highlighting the indissoluble link between the unity and diversity of all that is human.

The education of the future must therefore be based on the principle of what is universal, thus situating man, contextualizing through the questions: who are we? where are we? where did we come from? and where are we going? These questions emphatically lead man to question his position in the world. In view of this, Morin's thinking must take into account that man is animated by three prominent circuits that are fundamental to his life as an active subject:

Circuit 1

brain/mind/culture;

Circuit 1

reason/affection/pursuit;

Circuit 1

individual/society/species.

These circuits interact, which means the joint development of individual autonomies. The challenge for the education of the future in these terms is to manage the process of human formation by investing in harmony between the idea of unity and the idea of diversity, where one does not annihilate the other.

III.IV Teaching earthly identity

> They need to understand both the human condition in the world and the condition of the human world, which, in the course of modern

33

The question of earthly identity is a very serious one. It is discussed by Morin, who emphasizes that education has ignored the condition of a planetary human race throughout the ages, including now. The tendency in this century's planetary era is to grow to recognize planetary identity and convert it into educational principles. We need to learn that we are multidimensional: as well as being cultural beings, we are also natural, physical, psychic, mythical and imaginary.

According to Morin (2000), it is necessary to show that all the continents of the world, all parts of the planet are interconnected in a planetary feeling. Although there is diversity, humanity is one and adds complexity. There is a planetary crisis, left over from the last century, of human detachment, isolation and fragmentation. It is necessary for education to find the complex in the midst of the crisis, which will show that all people are faced with the same problems.

The 20th century left regenerative counter-currents in its wake.

1. The ecological backlash which, with the growth of degradation and the emergence of technical/industrial catastrophes, is only set to increase;
2. The qualitative counter-current which, in reaction to the invasion of the quantitative and generalized standardization, strives for quality in all fields, starting with the quality of life;
3. The counter-current of resistance to the purely utilitarian prosaic life, which manifests itself in the search for the poetic life, dedicated to love, admiration, passion and celebration;
4. The counter-current of resistance is the primacy of standardized consumption, which manifests itself in two opposing ways: one, through the pursuit of lived intensity (consumerism); the other, through the pursuit of frugality and temperance (minimalism);
5. The counter-current, still timid, of emancipation from the omnipresent tyranny of money, which is seeking to be counterbalanced by human and supportive relationships, pushing back the reign of profit;
6. The counter-current, also timid, which, in reaction to the unleashing of violence, nurtures ethics of pacification of souls and minds.

In opposition to the counter-currents, Morin highlights the need to inscribe ourselves in a universe of consciousnesses:

> **Anthropological awareness**, which recognizes unity in diversity. **Ecological** awareness, that is, the awareness of inhabiting, with all mortal beings, the same living sphere (biosphere): recognizing our consubstantial union with the biosphere leads to abandoning the Promethean dream of domination of the universe in order to nurture the aspiration of coexistence on Earth. **Earthly civic awareness**, i.e. of responsibility and solidarity with the children of the Earth. **The spiritual awareness of** the human condition that arises from the complex exercise of thought and which allows us, at the same time, to criticize each other and to self-criticize and understand each other. (MORIN 2000, p.73)

We are back to reflecting on the human condition as the object of understanding the education of the future. To know the human being is to situate him in the universe [...]. We must recognize our double rootedness in the physical cosmos and in the living sphere and, at the same time, our properly human rootlessness

> [...] We are original to the cosmos, to nature, to life, but because of our humanity, our culture, our mind, our consciousness, we have become strangers to this cosmos, which seems secretly intimate to us. [...] The human is a being who is at once fully biological and fully cultural, who carries within himself the original uniduality. (MORIN, 2004, p.51-52)

III.V. Facing uncertainty

> Knowledge is therefore an uncertain adventure that permanently carries the risk of illusion and error. However, it is in doctrinaire, dogmatic and intolerant certainties that the worst illusions are to be found; on the contrary, awareness of the uncertain nature of the cognitive act constitutes the opportunity to arrive at pertinent knowledge (MORIN 2000, p. 86).

One of the legacies of the development of science is the fact that it has placed humanity on a foundation of evidence, namely certainty. However, this is also a territory of countless uncertainties. So what Morin postulates is the need for teaching that includes uncertainty. Since it is extremely important to learn to walk in a terrain where certainties are only constellations in a galaxy of uncertainty.

In the author's opinion, the education of the future needs to connect with uncertainties, and he lists four principles:

1. A principle of cerebral-mental uncertainty, which stems from the process of translation/reconstruction inherent in all knowledge.
2. A principle of logical uncertainty: as Pascal said very clearly, "Neither contradiction is a sign of falsehood, nor non-contradiction a sign of truth."
3. A principle of rational uncertainty, since rationality, if it doesn't maintain vigilant self-criticism, falls into rationalization.
4. A principle of psychological uncertainty: it is impossible to be totally aware of what is going on in the machinery of our mind, which always retains something fundamentally unconscious. (MORIN 2000, p. 82)

In view of this, the real is not even legible, so it is extremely important to establish principles for interpreting immediate reality. This is because knowledge is often impregnated with doctrinal ideas, intolerant dogmatic beliefs, where the worst errors and illusions occur, in the author's opinion, (Morin, 2000).

In this context, he postpones the notion of the ecology of action, which takes into account the possibilities of the unexpected and the unforeseen while dealing with drifting consciousnesses.

Morin (2000) postulates that we have to teach the principle of uncertainty, in which scientific knowledge is never an absolute producer of certainties. On the contrary, everything created by man is riddled with the idea of uncertainty. Uncertainty can command the advance of knowledge and culture. This idea needs to be incorporated into the teaching of physics, chemistry, history, geography, languages and philosophy.

III.VI. Teaching comprehension

> Communication is triumphant, the planet is criss-crossed by networks, faxes, cell phones, modems and the Internet. However, misunderstanding remains widespread. Undoubtedly, there are important and multiple advances in understanding, but the advance of incomprehension seems even greater. (MORIN 2000, p. 90)

Still on the subject of technocracy, Morin considers the need for understanding to be both the means and the end of human communication. However, what we see is that 21st century education is still absent from this compression, even though it needs to be returned to all age groups. Teaching understanding is therefore the emblem that the education of the future must adopt, which undoubtedly calls for a reform of human thought.

We also return to the principle of the planetary era, in which human beings are connected as a species, are in fact close to each other, whether strangers or absentees, for humanity's sake. What is clear from Morin's reflection is that man is still in a rudimentary situation of incomprehension that needs to be broken down, transformed and deconceptualized. Hence the importance of understanding the roots of this incomprehension and the relationship between cause and effect in its modalities.

According to Morin (2000), there are two forms of understanding: intellectual or objective understanding and intersubjective human understanding. In this context, it seeks to perceive the intelligible in a connection between the text and its context, the parts and the whole, the multiple and the one. From this perspective, the process of understanding necessarily involves a process of empathy, identification and projection. Always intersubjective, understanding requires openness, sympathy and generosity. This highlights the fact that there are many external obstacles to intellectual understanding:

a. The "noise" that interferes with the transmission of information, creating misunderstanding or non-understanding;
b. The polysemy of a notion which, when stated in one sense, is understood in another;
c. There is ignorance of the other person's rites and customs, especially courtesy rites, which can lead to unconsciously offending or disqualifying oneself in the eyes of the other (cultural diversity);
d. There is a lack of understanding of the imperative values propagated within another culture - respect for the elderly, religious beliefs, the unconditional obedience of children, or, on the contrary, in our society, the cult of the individual and respect for freedoms;
e. There is a lack of understanding of the ethical imperatives proper to a culture, the imperative of revenge in tribal societies, the imperative of the law in evolved societies;
f. It is impossible, as a worldview, to understand the ideas and arguments of

another worldview, just as it is impossible for an ideology/philosophy to understand another ideology/philosophy;

g. Finally, there is the impossibility of understanding one mental structure in relation to another.

In the field of ethics, it is the art of living that gives stability to understanding. The ethics of understanding asks us to understand incomprehension. The teaching of understanding must enable harmony and acceptance of the new in societies in an open and democratic regime. In this way, there will be a more dialogic path between cultures, since understanding is also acceptance of the other's culture. So the challenge for education in the future is to teach for a minimum feeling of generalized understanding in a planetary era.

III.VII. The ethics of the human race

> The triad of individual/society/species are not only inseparable, but co-producers of each other. Each of these terms is both the means and the end of the others. You can't absolutize any of them and make one the supreme end of the triad; the triad is, in itself, its own end. (MORIN 2000, p. 102)

In this chapter, Morin brings the triad of individual/society/species into dialog, as elements of the education of the future which, as he postulates, are elements that need to be understood together and not dissociated. The central idea of the author's thinking is that the human being itself is a set of dimensions to be understood. Thus, the human gender means the development of individual autonomy, but in a collective context. Here he takes up the idea that although different, and geographically separated, human beings are equal as a species and act in community.

From this perspective, Morim presents an element that moves the individual/society/species triad, namely: Anthropo-ethics presupposes the conscious and enlightened decision to: assume the human condition individual/society/species in the complexity of our being; to reach humanity in ourselves in our personal consciousness, to assume human destiny in its antinomies and fullness.

The anthropological mission of the millennium is thus established through anthropo-ethics. It is clear that this mission changes human thinking and action. What the author emphasizes is that there must be a collective and human effort to work towards the humanization of humanity, since man has lost this sensitivity.

What becomes clear through anthropo-ethical intervention is the need to recover humanity's ternary character. According to Morin, humanity needs to be more human and this will always be the result of personal effort, but it has an influence on the collective. In this sense, individual/species ethics requires the mutual control of society by the individual and of the individual by society, in other words, democracy; individual/species ethics calls for earthly citizenship in the 21st century.

Analyzing Morin's propositions on anthropo-ethics, it can be said that it has two well-defined objectives in the education of the future: the first concerns control between individuals and society, this control being mutually governed by the principle of democracy. Perhaps this is in fact the pillar of human relations in a planetary society. The second concerns the process of education's contribution to raising awareness of our "Earth-Patria", hence the need for man to exercise earthly citizenship and not causal and ephemeral personal citizenship.

So there is no way of creating a single consciousness, and this would be neither ethical nor humane, but we do have arrows that show ways of opening doors to a better future. Morin's work does indeed present arrows on the road to the education of the future, but deciding to follow them belongs to the present. The education of the future begins now where we are, in the era we are in, with the remnants we still have of humanity.

Morin (2000) presents ideas that can help educators redefine their position in educational institutions in their relations with students, the curriculum, subjects and assessment. In summary, the author presents arrows along the way, through the following axes: Considering constant errors and illusions in conceptions; Building relevant knowledge; Re-learning our own human condition; Recognizing our earthly identity; Facing the constant uncertainties in scientific knowledge; Teaching understanding through dialogue and understanding; Discussing and exercising ethics.

In the light of the analysis carried out on Edgar Morin's "The Seven Knowledges Necessary for Future Education", I come to the conclusion, and when I use the term conclusion it is in the sense of understanding not exhausting ideas, that its main purpose would be to reform teaching and human thinking.

The book provides rich contributions for carrying out work that starts from the complex and not from the fragment. In practice, it proposes a redefinition of curricula that integrates knowledge and provides training and support for a new type of teacher. However, it should be understood that the thought of complexity put forward by Morin (2000) does not aim to disseminate an idea of disciplinary abolition; on the contrary, it encourages the possibility of introducing the new, the different and the complex.

In this way, it can be said that the education of the future requires greater integration of curricula in order to interconnect knowledge, while this possibility lies in the reorganization of pedagogical thinking. We need to move away from the casual and trivial. To enter the universal, planetary. The education of the future takes place through planetary practices and knowledge.

CONCLUSION

Morin's maxim challenges us to replace thinking that isolates and separates with thinking that distinguishes and unites, and this implies renewing the curriculum by incorporating everyday problems and interconnecting knowledge.

In his method, he defends the interconnection of all knowledge, combats the reductionism installed in our society and values the complex. Morin (2000) quoted by Andrade (2017, p.24) postulates that "human beings tend to push aside everything that is or seems complicated". Hence the urgent need to change the way we think. Since this is the only way to understand that simplification does not express the unity and diversity present in the whole.

It is important to emphasize that Morin does not condemn specialization, but rather the loss of overview. For him, the classroom is a complex phenomenon that is home to a diversity of moods, cultures, social and economic classes, feelings. It is a heterogeneous space and therefore the ideal place to initiate the mentality he preaches.

> "Morin's ideas for the classroom have everything to do with the current imperative for school to make sense to students. You learn more geography and history on a trip because it's easier to understand how much the content is part of a context" (Nova Escola, August/September 2002).

Thus, "the whole benefits teaching because the student looks for connections in order to understand. It is only when they leave the discipline and are able to contextualize that they see the connection with life" (MORIN 2000, apud de ANDRADE 2017, p.25). The school, like society, has fragmented in search of specialization. First, it divided knowledge into areas and prioritized certain contents within them. For Morin's ideas to be implemented, it is necessary to reform this structure, a complicated task.

REFERENCES

ANDRADE. Paulo Marcos Ferreira Andrade. **Pedagogies of the MST: A Possible Dream**! 1ª Ed. Novas Edigoes Academicas; SP, 2017.

Morin, Edgar - **The Seven Knowledges Necessary for the Education of the Future**, 3rd edition - Sao Paulo - Cortez; Brasilia, DF: UNESCO, 2001

NOVA ESCOLA. The Teacher's Magazine. Ed. August/September 2000

CHAPTER IV

Teacher Training in the Context of the Pedagogical Proposals of Rudolf Steiner (Waldorf Pedagogy), Maria Montessori and the Escola da Ponte Experience by Evelaine C. dos Santos

Evelaine Cruz dos Santos is carrying out her doctoral thesis with the aim of understanding the training processes of teachers who work in the pedagogical proposal of Waldorf Pedagogy, Maria Montessori and the Escola Da Ponte Experience. Her aim is therefore to outline the aspects of training in these three pedagogical strands, thus carrying out an effective study that highlights the elements that corroborate teaching praxis.

The work follows the perspective of elucidating the theoretical assumptions that surround these pedagogies, i.e. what the way of doing each of them implies. Thus, in Waldorf Pedagogy, the reflection goes through the theoretical-philosophical-methodological assumptions that underpin the teaching support, highlighting issues of self-knowledge, the arts and the class teacher. In the Montessori method, the dialogue focuses on the theoretical-philosophical-methodological assumptions that underpin it and their consequent updating, the importance of practice/internship and self-knowledge. In the Escola da Ponte experience, reflection stood out in school-centered training, with an emphasis on the study circle. So the key question that drives the research is: How does the process of training teachers to work in the Waldorf, Montessori and Escola da Ponte pedagogical proposals take place?

For a better understanding of this work, it is important to emphasize that pedagogy is understood as the way of doing teaching within the context of cognitive construction, be it formal or informal. According to the author, her research makes an effective contribution to "discussions on the training of teachers who teach mathematics, pointing in particular to the teacher's inner training through self-knowledge, an aspect that is considered in the Waldorf and Montessori proposals" (SANTOS 2015, p.05).

I. Teacher Training in Waldorf Pedagogy

As Santos (2015) points out, Waldorf pedagogy was born after the First World War, a time of profound transformation in Germany. It was proposed by "the philosopher, educator and artist Rudolf Joseph Lorenz Steiner, who was born on February 27, 1861 in the town of Kraljevec, which at that time was on the border between Hungary and Austria and is now part of Croatia" (MESQUITA 2011, p.10). The axis of its teaching proposal is based on an understanding of the legal-administrative, cultural-spiritual and economic triad, which has an intimate relationship with the ideals of equality, liberty and fraternity set out in the French Revolution.

According to studies by Santos (2015), the first Waldorf School was founded in 1919 in Germany and its aim was to offer primary education to proletarian children. However, according to some scholars, it was still on the fringes of pedagogical reform, even though it was already considered a reform school at that time.

Waldorf pedagogy emphasizes the integral development of the child, alluding to the aspects of the physical body, action, feeling, imagination, social and spiritual relationships. In order to understand Waldorf pedagogy properly, it is necessary to understand anthroposophy, since it is the practice of anthroposophy.

So we have the following definition:

> Anthroposophy is a path of knowledge that aims to bring the spiritual of the human entity to the spiritual of the universe. It appears in the human being as a need of the heart and feeling, and must find its justification in the fact that it can provide satisfaction for this need. Anthroposophy can only be recognized by a person who finds in it what, based on their sensitivity, they must seek. Therefore, only people who feel the need to ask questions about the essence of the human being and the universe, just as one feels hunger and thirst, can be anthroposophists (STEINER, 1923, GA 306).

For Steiner (2006b),

According to Santos (2015), anthroposophy is a spiritual science whose universe of principles encompasses all the manifestations of life. The author's propositions highlight the soul as an inseparable element in the study of human development. Thus, the understanding is that the origin of all knowledge is in the spirit of man, while knowledge is a reflection of man himself "and the latter recognizes himself in the former" (p.13).

From this context, it is clear that Waldorf pedagogy has a conception of man in a harmonious relationship between the physical-animal-spiritual, the principle on which the educational action is based. It starts from the hypothesis that the human being is not determined exclusively by inheritance and the environment, but also by the response that he or she is capable of making to the impressions he or she receives from within. It considers that man is born with a potential of predispositions and capacities which, throughout his life, struggle to develop (MIZOGUCHI, 2006).

Based on this assumption, children's learning is divided into developments that last for periods called septennia. For a better understanding, Santos (2015) explains the septennia as follows:

1. **In the first seventy** years (ages 0 to 7), the child attends nursery school. "This period is characterized by the structuring of the physical body and the child is permeable to all the influences of the environment, as if it were a large sensory organ (...)" (SANTOS 2015, p.29);

2. **The second seventieth** year (7 to 14 years) sees the child attend elementary school. In this period, the astral body takes the lead, and with this all development is linked to feelings, emotions, providing a basis for the child's psychological maturation (SANTOS 2015, p.29);

3. **The third septennium** is developed by young people when they are in secondary school (14 to 21 years old) and is considered the basis for social maturation. In this period, the self becomes autonomous, which means the full development of mental and moral faculties (SANTOS

2015, p.30);

The third term would then be the consolidation of intellectuality. In addition to the basic subjects of the curriculum, activities related to the arts and aesthetics are developed. Waldorf pedagogy also works with the concept of the class teacher, who would be the professional with direct responsibility for the same class for a period of eight years.

> There is basically no education at any level other than self-education. [...] All education is self-education and we, as teachers and educators, are really just the environment of the child educating itself. We must create the most suitable environment for the child to educate himself together with us, in the way that he needs to educate himself through his inner destiny (STEINER, 1923 quoted by MESQUITA 2011, p.18).

According to Santos (2015), although the Waldorf school has a pedagogical organization that enables the development of the human person in all its dimensions, the challenge of teacher training still persists in its dynamics. Thus, the object of the research for his thesis is established on this challenge, namely teacher training.

I.I. Santos' considerations on Waldorf teacher training

The first reflection postulated by santos (20150) is in fact about the criticisms of Waldorf pedagogy inferred in academia, although they make no mention of its presuppositions in training courses. Her research found that several studies have been carried out by researchers on Waldorf pedagogy, but that very little attention has been paid to the issue of teacher training.

In this context, the author states that:

> Little attention has been paid to Waldorf teacher training. In the search for research carried out in the period from 1980 to 2011 on the subject of Waldorf teacher training, we can highlight Verilda Speridiao Kluth's dissertation (1997), Alduino Mazzone's doctorate (1999), Mila Carvalho Gomes's course completion work (2008), Salles's dissertation (2010), the research by the Fundo Unico de Bolsas - FUB35 (2012) and the Rewa group (2012) (SANTOS 2015, p. 33).

In these terms, Santos (2015) continues by asserting that the work carried out by Kluth, although it focused on the participation of Waldorf teachers in refresher courses, did not in fact address teacher training, "but rather the analysis of the subject-mathematical encounter, using Merleau-Ponty as a theoretical contribution" (p.35). Gomes and Mendes (2008) made inferences from the perspective of understanding "the role of the Waldorf teacher based on their training and pedagogical practice, detailing aspects of man in the phases proposed by Anthroposophy" (p.35). This work is substantiated by what the teachers say as they prepare their work. In my opinion, this is already an important step towards understanding teaching.

In light of this, Santos (2015) cites "the work of Salles (2010) aimed to analyze how teachers of the 1st cycle of elementary school, who participated in the Dom da Palavra Project" (p. 36), she also goes on to explain the research of Mazzone (1999) and an analysis of the Waldorf teacher in Australia, "which covered various facets of Waldorf teacher training, however, it did not deal with how mathematics teaching is approached in this training (p.36). In the author's view, although the research cited is relevant, there has not yet been an approximation between teacher training and mathematics teaching.

Santos (2015) then highlights the characteristics of the Waldorf elementary school teacher in three interconnected areas:

1. Deep knowledge of the human being. This involves knowledge of anthroposophy;
2. Love as the basis of social behavior towards students;
3. Artistic qualities, (LANZ 2005, p. 86 apud de SANTOS 2015, p.38).

In order to work effectively in Waldorf pedagogy, the teacher must develop techniques linked to the arts, since the arts underpin the Waldorf pedagogy. [...] In every class, the teacher tries to develop the materials in an artistic way and the same subject is approached differently each time, always according to the students' stage of development (MAIA, 2004, p. 13 quoted by SANTOS 2015, p.38). According to the author, the Waldorf school carries out its pedagogical work with three types of teachers, which she identifies as follows:

1. Class teachers;
2. Specialist or subject teachers;

3. Assistant professors.

When starting a career in a Waldorf school, the teacher must first take responsibility for a class, where "the focus is on the relationship between teacher and students, on the development of the class as a whole, awakening skills, abilities and knowledge" (SANTOS 2015, p. 40).

Gomes and Mendes (2008, p.62) quoted by Santos (2015, p.40), state that:

> When starting their professional career, Waldorf teachers are faced with the challenge of accompanying a class for eight years. They will be intensely present in the essential learning years of a human being's life. He or she will be the guiding light, the example, the safe haven that the child will look to. They will have the responsibility of promoting the individual's education in the conquest of their freedom and autonomy, guiding them towards a correct integration into social life.

Santos (2015) adds that:

> From a legal point of view, to work as a class teacher in a Waldorf school in Brazil, you need to have a degree in Pedagogy, which qualifies you to teach grades 1 to 5. For the 6th to 8th grades, there is no legislation to support the class teacher, as specialists come in and the class teacher doesn't have a degree in all the subjects taught at this level. Waldorf schools usually require a foundation course in Waldorf Pedagogy42 which complements the pedagogical training in general degrees. In São Paulo, the Waldorf Teacher Training Course at the Rudolf Steiner Waldorf School has been recognized and qualifies the class teacher. (SANTOS 2015, p.42)

As a detriment to this, the school's memory tells us that on the occasion of the inauguration of the first school, an intensive pedagogical training course for teachers was given by Rudolf Steiner for three weeks. In Brazil, the first Waldorf Pedagogy seminar was held because of the need to renew the teaching staff. According to Santos (2015), the seminar was founded by Rudolf and Mariane Lanz, the school's founding couple, in 1970.

In these terms, training, postgraduate, qualification and foundation courses in Waldorf pedagogy are currently offered in training centers in Sao Paulo, Santa Catarina, Ceara and Pernambuco. When the author joined one of the Waldorf training courses, she was able to perceive the training of teachers in locus, as she outlined the following structure:

1. **First theme** - introductory presentation to the themes of Trimembragao and Quadrimembragao and Watercolor Painting (duration: one weekend).
2. **Second theme** - introductory presentation to the themes of the Twelve Senses and the Sevens, and Music (duration: one weekend).
3. **Third theme** - introductory presentation to the themes of The Pedagogical Question as a Social Question and Cosmogony and the Design of Forms (duration: one weekend) (SANTOS 2015, p.56)

According to the author, a second stage took place, which she describes as lasting 362 hours under the theme "Human biography and its seventies":

1. **First theme** - The trimembered man - body, soul, spirit (duration: one week);
2. **Second theme** - The fourfold man - the bodies: physical, etheric, astral and I (duration: four weekends);
3.   **Third theme** - The twelve senses (duration: one week);
4. **Fourth theme** - Human biography and its seventies (duration: four weekends);
5. **Fifth theme** - The pedagogical question as a social question (duration: one week)
6. **Sixth theme** - Anthroposophical cosmogony (duration: four weekends) SANTOS 2015, p.57).

In addition to the theoretical contributions that were studied, the course worked on the concepts of the arts, social awareness and an introduction to anthroposophy. The author reports that, in interviews, the teachers who took part in the course argued that they found it difficult to understand the Waldorf Pedagogy proposal.

Subsequently, the third stage was carried out with a workload of 228 hours. Santos (2015) reports that:

> The themes covered in the theoretical part were: The content and practice of annual festivals; The first three years of life; The twelve senses in Seventh Grade Education; The rhythmic wheel; Observing children; Fairy tales; Play and toys; Reading children's drawings; and Working with parents. (SANTOS 2015, p.61)

The fourth stage lasted one year, with a workload of 228 hours, followed by the fifth stage where the course ended.

> The topics covered in the theoretical part were: The first three school years and the 1st rubric, the 4th and 5th school years, the 6th, 7th and 8th school years and the 2nd rubric, Methodology and Didactics, Methodology and Didactics in Science Teaching, Goetheanistic Observation and Child Observation, (SANTOS 2015, p.63).

The fourth stage also focused on "Mathematics and Geometry" according to Santos (2015, p.65) "the seminar that offers a general formation for the Waldorf proposal". There was a stage "where, in the Anthropological/Anthroposophical Foundation stage, Geometry appeared separately from Mathematics, with a greater focus on Geometry" (p.66) and another "dedicated to Kindergarten, Mathematics and Geometry were barely mentioned" (p.68). In the "Conclusion stage, Mathematics and Geometry were not mentioned. In the Elementary School stage, there was a strong focus on Drawing Shapes" (p.69).

Santos (2015) points out that training also

> Mathematical content from the 1st to the 5th grade was covered, such as the qualitative introduction to numbers, tables, mental calculation, prime numbers, Roman numerals, maximum common divisor, among others; content that goes against current trends in Mathematics Education. (...) Mathematics teaching in the 6th, 7th and 8th grades was little commented on, being brought up more in terms of cumulation (prescribing what to work on with the students). We can see the precariousness of training for teaching mathematics in Primary II (5th to 9th grade). Perhaps this justifies the fact that many class teachers are not taking on Mathematics in Primary II, leaving it to the specialist. (SANTOS 2015, p.69)

According to the author, the interdisciplinary processes involving mathematics, history and handicrafts were also evident in the training. The teacher responsible for the training also took the initiative to make it possible to be enchanted by mathematics, although according to Santos (2015) she was not concerned with the didactics of mathematics.

In summary, the author indicates that the course strengthened the cultural range of the participants, as well as their interaction and engagement. During the classes, many things were left open, since there is a view that there are no recipes and that in the Waldorf proposal there is no method, (p.72).

Santos (2015, p.75) assures that:

> As for the trainers' backgrounds, most of them have a great deal of experience as teachers in Waldorf schools, as well as being active in the Anthroposophical movement. Most of them don't have a postgraduate degree, which makes it difficult for the MEC to recognize them as higher education courses.

The author carried out exploratory research with the participants in order to understand their assessment of the course. While some said they had difficulty understanding the proposal and even assimilating its application, while others identified their artistic difficulties, in general terms the participants considered the training to be excellent, stating that it was indeed worth taking part in Waldorf teacher training.

The author considers that there were difficulties in understanding the concepts of Anthroposophy, while another participant "sees the study of Anthroposophy as an exercise in self-education and seems to have had no difficulties" (SANTOS 2015, p.79). "The Waldorf proposal develops its curriculum based on the development of the human being and takes anthroposophy as its basis. It requires specific training and this implies a new attitude for the professional" (p.81).

The author points out that in order to work with Waldorf pedagogy, teachers need to understand anthroposophy, since it is the foundation of the pedagogical proposal. Therefore, its foundations must be considered as a basic and necessary principle. Because "Waldorf schools require a group of anthroposophical educators in their professional staff" (SANTOS 2015, p.81). In these terms, the main elements of teacher training in Waldorf pedagogy are anthroposophy, and the reflection within the seven years of self-knowledge and the class teacher as the main pedagogical realization in the school.

Her research concludes that "it can be deduced that all Waldorf teacher training is based on the theoretical-philosophical-methodological assumptions of the pedagogical proposal itself, which are based on Rudolf Steiner's

Anthroposophy (SANTOS 2015, p.81).

In short, teacher training works to ensure that teachers develop the following qualities in order to consolidate a praxis that is coherent with the principles set out:A deep understanding of the human being;

1. Mastery of Anthroposophy;

2. Perception of the development of the human being through the seventies;

3. Love as the basis of social behavior;

4. Artistic qualities, in terms of malleability, fantasy and creativity, seeing each lesson as a work of art;

5. Master your own temperament and language, avoiding abstractions and speaking concretely and imaginatively;

6. Struggling to deal with problems and situations whose scope would normally have eluded you;

7. Sensitivity and the ability to recognize his work in the students themselves, discovering where questions and doubts arise in the souls of his students (LANZ 1979 quoted by Mesquita 2011, p.30),

II.   The Montessori method and teacher training

Santos (2015) begins his reflection on the Montessori method by giving a brief biographical context of the Italian doctor, which is transcribed in full in this document:

> She was born on August 31, 1870, in the small town of Chiaravalle, Italy. Her family was religious and conservative. In 1875, her family moved to Rome. Opposing her family and overcoming the resistance of the time, in 1892 Montesssori attended medical school. According to Kramer (1976), his entry into the Faculty was authorized by Pope Leo XIII. In 1896, she graduated from the University of Rome with a thesis in psychiatry and was involved in the feminist movement (MMONTESSORI, 2010). During her doctorate in medicine at the University of Rome, Montessori became acquainted with the work of Edouard Seguin, who was involved in the treatment and education of the "abnormal" (the term used at the time for children with disabilities). For 12 years, she worked as an assistant professor at the university's Psychiatric Clinic, dedicating herself to the treatment of children with disabilities, where she concluded that the issue of these children was more pedagogical than medical and thus turned to the application of a

Due to her experience working with children, Maria Montessori was appointed to direct the work of early childhood education at the first Children's House, which opened in 1907 (SANTOS 2015).

Schools based on the Montessori method offer a unique dynamic where pedagogical work takes place through excellent organization and scientific materials prepared to meet children's rhythms.

Santos (2015) points out that the groupings are made by subdividing the age groups as follows: "Early Childhood Education generally has one grouping with children aged 0 to 3 and another with children aged 3 to 6. In elementary school, there is a grouping with children aged 6 to 9, another with children aged 9 to 12 and another with young people aged 12 to 15" (p.83).

This dynamic of divisions postponed by Maria Montessori was a great innovation since the idea of mixed classrooms was in fact a novelty, since there were only classes formed with children of the same age. Mixed classrooms thus developed a method based on cooperation, in which older pupils contributed to the learning of younger ones.

From this perspective, the classroom becomes an environment of freedom, where those involved choose their "occupations, work, talk and even do nothing (LENVAL, n.d., p. 93 cited by SANTOS 2015, p.84). This is a space that aims not to influence the child but to meet their needs, as it must, by rule, offer the following conditions:

1° Promote children's scientific knowledge;

2° Establish an environment of freedom and respect for the child;

3° The educational environment must be aesthetically pleasing;

4° The child must be active;

5° Children must be able to educate themselves;

6° The child must correct him/herself, and the teacher is not responsible for correcting him/her;

7° The teacher must essentially observe (MARCHIORI, 2009 cited by SANTOS 2015, p.84).

The activities are carried out mainly with the hands, and the method explores the senses through movements based on the principle of the construction of humanity. The idea is that motor activities enable children to perceive themselves while also perceiving the movement they make. The early childhood education curriculum is made up of areas of knowledge that involve: "Mathematics and Geometry, Language, Social Studies (History and Geography), Science, Arts, Music, Motor Development, Practical Life and Sensory Education; which are grouped into three broad categories: motor education, sensory education and intellectual education" (SANTOS 2015, p.85).

The Montessori method works on the principle of demonstration, where an adult shows the experience and then the child, where the concepts of memorization and verbalization are developed. The dynamic completely excludes the effects of reward and punishment, and emphasizes cosmic education, where the teacher helps to develop the child's own strength of being. In other words, the abilities are already there and must be stimulated by adults.

According to the author:

> At the Copenhagen Conference in 1937, Maria stated: -the great urgency today is the creation of a Science of Peace and the education of men for peacell76. On that occasion, she proposed organizing efforts towards a scientific education capable of contributing to security and progress. She said that education should reach the same level of excellence as the progress of science and that these efforts should focus on the child as the maker of the New Man. -The child is living proof that men can change and improve from birtholl (OMB, 2011b cited by SANTOS 2015, p. 87).

Montessori thought postulated the existence of an inner force in the structure of the child itself, while psychic acquisitions are realizations of the sensory periods of each child. Thus, through the concentration of developed activities, Montessori creates the principle of normalization. "Normalization comes from concentrating on a task. To this end, there must be motives in the environment that are capable of provoking this attention; objects must be used according to the purpose for which they were built, in other words, that lead to a mental order" (MONTESSORI S.D quoted by SANTOS 2015, p.88).

According to Santos (2015), the education of children using the Montessori method differs from the education of adults in five important ways: the environment and the adult, grouping, autonomy, independence and concentration. Montessori therefore proposed a curriculum based on: practical life, sensory education and motor development, Mathematics and Geometry,

Language, Science and Social Studies (History and Geography, Arts and Music Education.

With regard to mathematics and geometry in early childhood education, Montessori (2006) notes that "human intelligence is a mathematical intelligence, and therefore education and mathematical development enable us to understand and participate in the progress of our time" (SANTOS 2015, p.92). In this way, it creates materials that stimulate the mind as it is designed to operate accurately.

In this context, some important aspects of the Montessori mathematical notion are evident, which are set out here by Lenval, (n.d, 191 cited by santos 2015, p. 93) and which we make a point of transcribing here in full:

1.   **First fundamental principle**: counting and comparing dimensions or quantities in relation to a unit of measurement. The unit chosen is arbitrary. If this is my unit, any other quantity will be a multiple of this unit .

2.   **Second fundamental point:** digits are signs that indicate relationships between dimensions and quantities. If this □ equals 1, then this equals 5 and this 4.

3.   **The third fundamental rule:** always count to nine. At ten, you start again. Nine units, nine tens, nine hundreds, etc.

4.   **Fourth fundamental notion**: each of the four operations has a specific role. It represents an action, or let's say a displacement of matter in space.

In this context, Santos (2015) postulates that "Mathematics in Maria Montessori's pedagogical proposal currently takes into account new research into the human brain" (p. 93).

The Montessori method arrived in Brazil in 1957 from the French side and in 1958 the first teacher training took place with Pierre Faure. This training later became a specialization course for teachers of the Montessori system, which was considered the first official teacher training course. In this context, the "Brazilian Association of Montessori Education (ABEM) was founded in 1973, at the request of Mario M. Montessori, with the aim of spreading his mother's thoughts" (SANTOS 2015, p. 106).

Although the author points out that she has not found any Brazilian theses that discuss teacher training or that highlight the teacher's task from a Montessori perspective. Santos (2015) infers that in Montessori pedagogy, educational praxis is not based on transmitting knowledge, but on sharing it in a given environment.

In this way, the adult plays a fundamental role in mediating learning situations, while not limiting the conditions, times and possibilities for the child to build their own paths. According to Montessori, the teacher must therefore be attentive to development, to what is happening:

> [...] His words, energy or severity are not necessary; what matters is an attentive spirit of observation, his vision in serving, interfering, withdrawing, keeping quiet, according to cases and needs. You must acquire a moral skill that no method has ever required before; a skill made up of calm, patience, charity and humility. It is virtues, not words, that are his greatest preparation (MONTESSORI, 1965, p. 144 quoted by SANTOS 2015, p. 111).

In short, "the educator should [...] teach little, observe a lot and guide the psychic activities of children as well as their physiological development. [...] (MONTESSORI, 1965, p. 156 quoted by SANTOS 2015, p. 111).

According to the author, Montessori pedagogy aims to train teachers based on the principle of self-training, i.e. the teacher must therefore seek out what is necessary to develop a cohesive and effective practice even before their academic training. Various training courses are offered to teachers, but only in Montessori schools.

However, the "OMB supports Montessori courses in Brazil, which are offered by three major centers: Centro de Estudos Montessori in Rio de Janeiro, Centro Educacional Menino Jesus in Florianopolis and The Center for Montessori Education in Sao Paulo, (SANTOS 2015, p. 114).

This is how Montessori teacher training is structured, using a dynamic approach focused on the education of children from 3 to 6 years old. The courses are coordinated by Paige and Marion, both American teachers.

The author notes that the training courses take place in two modules per semester, where students have to complete various procedural and conceptual targets, and are certified when they have completed 90% of the classes and completed a corresponding 240-hour internship. This is not a public course, but a private one with payment for each module.

The teachers' training is theoretically based on Maria Montessori's contributions and also on the thinking of authors who support her practice. The course has 14 trainers who work as follows: "guest lecturer (2), additional instructor (6), course teacher (4), course coordinator (2). The guest lecturer gives very few lectures on the course" (SANTOS 2015, p. 124).

After studying the theory, the teachers visit the schools and carry out the internship, which is monitored by the course directorate.

With regard to mathematics, the training course had a special stage to give more emphasis to reinforced mathematics, since according to the course coordinators, the mathematics studied at university is very superficial. "The view of mathematics adopted by the teachers was explained in the theory lesson. In this lesson, the teacher posed questions and riddles about mathematics and several students tried to answer them and got involved" (SANTOS 2015, p. 133).

The Montessori method has received a lot of criticism at a national level, which according to Santos (2015) may be one of the causes that have hindered its diffusion in Brazil. Santos (2015, p.135) cites Monteiro (2009) who raises three hypotheses:

a. Teacher training (colleges don't work with Montessori ideas and people have to look for a specific course);
b. The physical structure of the Montessori school is very different from what is usual in Brazil;
c. The need for materials, which are expensive and not easy to find.

In this context, Santo points out that as the material is expensive, it can be made by the teachers themselves so that the method can be applied in the classroom without any loss of quality.

Santos (2015) concludes by evaluating the Montessori teacher training course and points to its organization and structure as positive. He goes on to say that "at the end of the course, the student has a solid specific training, as well as plenty of material for working in the classroom, because both the academic basis and the practice of the Montessori method were well covered throughout the course" (p.136).

I II.Understanding the dynamics of the Bridge School

We indicate the reflection with this epigraph used by santos (2015, p. 140) "the school I always dreamed of, without imagining it could exist ..." The author shows through her research that A escolada Ponte came into existence "in 1932, and is one of seven public schools in Vila das Aves, Portugal. Vila das Aves has an area of approximately 6 km² and 8,492 inhabitants117. It was elevated to town status on April 4, 1955, and until then was known as Sao Miguel das Aves" (ibid.).

The Ponte school has a revolutionary dynamic because it has rethought the teaching method and the conception of school, radicalizing traditional pedagogy. It is coordinated by Professor Jose Francisco de Almeida Pacheco, who postulates that it was an archipelago of solitudes that, in a systematic drift, allowed "transformations in communication structures and intensified collaboration between institutes and local educational agents" (PACHECO, 2012c cited by SANTOS 2015, p. 141).

The current school dynamic is based on the principle of bridging the gap. In other words, mediating autonomous learning, personal choices and common interests. It is from this perspective that "[...] the word bridge acts both as a designation of the school and as a metaphor. It evokes change. It will be a place through which, between, or across which one can pass from the possible to the necessary" (ESCOLA DA PONTE, 1996, p. 21 cited by Santos 2015, p. 141). Its dynamics categorically implied a change from disciplines to projects, but this also required a new physical structure that would actually cater for the changes that had taken place.

Whereas Pacheco considers:

> [...] This project includes an area that I call the discovery center, where
> we will share what we know. There are also small hexagonal niches for
> small groups and individual tasks. There are also wide avenues and
> some waterways where you can dip your toes in to chat, as well as a
> place to nap. The new information technologies must be spread out
> everywhere so that they can be democratically used by the community,
> which we have already achieved (MARANGON, 2004 apud de
> SANTOS 2015, p. 144).

This new physical format has enabled the consolidation of the school's three main values: freedom, responsibility and solidarity (SANTOS 2015). The school caters for elementary school pupils, with approximately 220 pupils. "The school's opening hours are from 8.45 a.m. to 3.45 p.m., with three study periods of approximately 1.50 minutes each, with a 30-minute break in the morning and a 1.10-minute lunch" (SANTOS 2015, p.145). The studies are directed by themes chosen in assembly each school year.

According to Santos (2015), the activities are developed in three nuclei that correspond to three cycles:

1- **The Initiation Nucleus** receives students from Early Childhood Education and the 1st cycle. Students leave this nucleus when they have acquired attitudes and skills that enable them to integrate into the school community and work autonomously (ESCOLA DA PONTE, 2012d apud de SANTOS 2015, p. 147).
2- **In the Consolidation core**, students solidify the skills they have acquired and try to achieve the nationally defined learning objectives (ESCOLA DA PONTE, 2012e apud de SANTOS 2015, p. 147).
3- **In the Deepening core**, students develop the skills of the second cycle of basic education and are expected to manage their time at school with complete autonomy (ESCOLA DA PONTE, 2012f apud de SANTOS 2015, p. 147).

According to Santos' report (2015), a student took part in the nucleus and, noticing difficulties with some of the content, asked for a direct lesson with the teacher. "He went around and asked who wanted to take part in the class. The teacher only allowed those who had already studied or were studying the subject to take part" (p.149). Given this context, the lesson was scheduled with the students who had already studied the subject and then the intervention was carried out with a focus on the difficulty. In this dynamic, the teacher is not the only holder of knowledge, but the student discovers the content before the lesson.

Santos (2015) reports that he witnessed a teacher carrying out a mathematics assessment with a girl, and from his perception the assessment action had more to do with dialog, of course with notes of the whole process. Realizing that the Ponte school doesn't work with the concept of proof. "At the end of each school year, in addition to the grading of each subject, the general attitudes/competences of the Project are also assessed. However, the Project and the projects are also continuously evaluated" (ESCOLA DA PONTE, 1996, p. 18 cited by SANTOS 2015, p. 152).

Educational scholars have postulated that the Bridge School has moved away from educational models where the teaching process is imposed, pointing to the Bridge School method as eclectic.

Delving a little deeper into the teaching of mathematics at Escola da Ponte, Santos (2015) found out on a visit to the school that "there is a proposal to solve a mathematical problem every two weeks. This work is carried out by the Logical-Mathematical Dimension specifically for each nucleus" (p.158).

> The ability to self-evaluate one's reasoning, procedures and answers is fundamental to learning mathematical concepts. **Solving problems, for example, is not possible without this aspect being resolved. On the other hand, the ability to compare answers with the initial question and check their consistency is extremely important [...] (MENSLIN, 2006b cited by SANTOS 2015, p. 158).**

The learning situations are supported by the use of manipulative materials, although Santos (2015) believes that the use of these resources should be intensified. While training and memorization are strengthened by the direct classroom system. In this way, the author posits that "although the school is working with projects, I think that the strategy of parallel work on mathematical content strengthens the view that it is possible to work in a transdisciplinary, multidisciplinary, interdisciplinary and also disciplinary way" (SANTOS 2015, p.159).

The most important thing, according to the author, is the understanding that at the Ponte school it is possible to work with all the possibilities for constructing knowledge.

III.I The Ponte School teacher training process

In the School of the Bridge, the teacher is called an educational advisor, which is in line with the thinking of the innovative school of the 20th century. These teachers "have degrees in a wide range of areas, some have specializations and others have master's degrees" (SANTOS 2015, p. 161). However, the author found that cooperative work is difficult, as it requires greater commitment on the part of the teachers.

Multiskilling is highly valued at Escola da Ponte, as the school understands the challenge it faces in developing a culture of cooperative work. The author says that in her fieldwork at the school, she noticed a positive aspect of the teacher/student relationship: "a math teacher interacting with 2nd/3rd cycle students. He didn't actually teach the students the content, but asked them questions to stimulate their thinking or answer their doubts" (SANTOS 2015, p. 162).

When it comes to the object of his research, namely teacher training, Santos (2015) states that there is still little scientific material that deals specifically with this subject.

Quoting Pacheco (2008), Santos (2015, p. 163) postulates that:

> teacher training is a continuous and participatory process, derived from and referenced to practice, a continuous process of action and critical reflection on action, because it is through critical reflection that forms of legitimization (of authority, or moral regulation, for example) are questioned.

It's clear from the research that teacher training is carried out in cycles of studies modeled on spontaneous practices. In 1978/79, the documentation center of Santo Tirso was founded, which set up ongoing training teams based on documents from the Ministry of Education. However, the group was disbanded in 1980.

On the basis of his on-site investigation, Santos (2015) considered the training of teachers at the Bridge School in three fundamental aspects:

a) Continuous training internship; "teacher preparation does not take place in a specific, systematized way, those who arrive learn from other colleagues, discussing day-to-day life, practicing what they are expected to be able to mediate with students" (MENSLIN 2006[a] apud de SANTOS 2015, p. 165);

And in a very specific way, training in mathematics has the following objectives:

1. To find out what's new in mathematics teaching and the management of materials related to the dimension;

   2. Draw up the problem for the fortnight;

3. Check the progress of the work in the Logical-Mathematical dimension in the three nuclei;

   4. Coordinating dimension meetings;

5. Carrying out their work in a spoken way that is interesting for everyone (SANTOS 2015, p. 168);

   b)   Continuous training at school (through team meetings, study circles, dimension meetings, etc.);

   c)   Fazer a Ponte Online Course: "Escola da Ponte has been offering a course called Fazer a Ponte since 2006. This is an online course led by Professor Wilson Azevedo, in which the school's teaching staff, parents, students and alumni take part" (SANTOS 2015, p. 170);

   Santos (2015) points out that there is in fact a continuing education project at the bridge school focused on mathematics education, but the author criticizes the non-participation of other teachers in this training since they work on the principle of polyvalence. However, she asserts that "the topics covered in the meeting of this dimension are in line with trends in Mathematics Education, providing a debate and a proposal for practice in line with emerging research in the area" (SANTOS 2015, p. 174).

IV.   Conclusion

At the end of the study, the author raises the possibilities of teacher training for mathematics teaching, considering Waldorf, Montessori and Escola da Ponte pedagogies.

His thesis shows very clearly how the training of mathematics teachers converges on certain curricular proposals and ideologies. According to Santos (2015, p. 188), "Mathematics appeared in these processes, focusing on the narratives of the researcher's experience of the course, and not on the narratives of the experience of the other subjects of the courses".

V. References

STEINER, Rudolf. **The Art of Education III**. Sao Paulo: Anthroposofica, 2000.

. **The Education of Children According to Spiritual Science**. Sao Paulo: Ed. Antroposofica, 3ª ed. 1996, GA 34.

. **A Pratica Pedagogica. Sao Paulo: Antroposofica**, 2000, GA 306.

. **The basic animic-spiritual forces of the art of education. Spiritual values in education and social life**. 1979, GA 305.

. **The healthy development of the human being**. Sao Paulo: FEWB, 1998.

49

. **Current spiritual life and education**. Dornach: Rudolf Steiner Verlag, 1986. GA 307.

MIZOGUCHI, Shigueyo Miyazaki. Rudolf Steiner and Waldorf pedagogy, **Colegao memoria da pedagogia - perspectivas para o novo milenio**, Rio de Janeiro, 2006.

MESQUITA, Marcos Roberto Linhares. **Waldorf Pedagogy: A holistic vision as a pedagogical approach** / Marcos Roberto Linhares Mesquita. 2011

SANTOS. Evelaine Cruz dos. **Teacher Training in the Context of the Pedagogical Proposals of Rudolf Steiner (Waldorf Pedagogy), Maria Montessori and the Bridge School Experience**. UNESP - Rio Claro Campus. Sao Paulo, 2015.

yes
I want morebooks!

Buy your books fast and straightforward online - at one of world's fastest growing online book stores! Environmentally sound due to Print-on-Demand technologies.

Buy your books online at
www.morebooks.shop

Kaufen Sie Ihre Bücher schnell und unkompliziert online – auf einer der am schnellsten wachsenden Buchhandelsplattformen weltweit! Dank Print-On-Demand umwelt- und ressourcenschonend produzi ert.

Bücher schneller online kaufen
www.morebooks.shop

Printed by Books on Demand GmbH, Norderstedt / Germany